Faith, hope and doubt in times of uncertainty

2008
THE **JAMES BACKHOUSE** LECTURE

Faith, hope and doubt in times of uncertainty

Combining the realms of scientific and spiritual inquiry

GEORGE ELLIS

About the author

George FR Ellis is Professor Emeritus of Applied Mathematics, University of Cape Town, South Africa, and previous Head of the Department of Applied Mathematics. His professional research work includes relativity theory and cosmology, complexity studies, the functioning of the brain and science policy. His academic positions have included being a university lecturer at Cambridge University and visiting professor at the University of Texas, University of Chicago, University of Hamburg, Boston University, University of Alberta, University of London (Queen Mary) and International School of Advanced Studies (SISSA), Trieste. He is a past president of the International Society of Relativity and Gravitation and of the Royal Society of South Africa, and is a Fellow of the Third World Academy of Science and of the Royal Society (London). He has been awarded the Star of South Africa Medal by President Nelson Mandela, the Order of Mapungubwe by President Thabo Mbeki, and the Templeton Prize (2004). His books include *The large scale structure of space time* with Stephen Hawking and *On the moral nature of the universe* with Nancey Murphy. He joined the Religious Society of Friends in 1974, and has been Chairman of Quaker Service, Western Cape, and of the Quaker Peace Centre Board. He has been Clerk of the Cape Western Monthly Meeting (CWMM) and of Southern African Yearly Meeting (1986-8). His web page is at: http://www.mth.uct.ac.za/~ellis/

Acknowledgments

I thank the Backhouse Lecture Committee and Judith Toronchuk for helpful comments that have greatly improved a previous draft of this talk.

Abstract

The rise of science over the past 300 years has led to an increasing series of attacks on religious faith, renewed with vigour in recent times, in particular by Richard Dawkins, Peter Atkins, Daniel Dennett and Viktor Stenger. Doubt about faith and religion has been strengthened by such attacks. So what are the intellectual resources and sources of spirituality that can sustain us in these times of uncertainty?

First we need to consider the nature and limits of science: what we learn from it, and what we cannot learn from it. Science discovers the physical context of life and the nature of physical causality. Reductionists tell us this is the only kind of causality there is (using the phrase 'nothing but' to emphasise their viewpoint), but this is wrong: there are other forms of causality in action in the world, in particular, whole-part causation and human intentionality. Non-reductionist views of science will take them into account, thus freeing us from the straightjacket of strong reductionist worldviews. The desire to free us from irrationality leads to the myth of pure rationality, suggesting pure reason alone is the best basis for a worthwhile life. But this is a completely inadequate understanding of causation on which to base a full life. Rationality, faith, hope and doubt as well as imagination, emotions and values are all important in a full understanding of human choices and decisions. They all interact with each other and are causally important in the real world. The key one is values, related to aesthetics and meaning (*telos*): this is what ultimately guides our choices and actions, and so shapes both individual lives and society.

Many important human endeavours and understandings of necessity remain outside the domain of science; these include the key issues of ethics, aesthetics, metaphysics, and meaning. I will discuss each of these briefly, and how they transcend scientific views. The source of values is a key point, and the various scientific proposals in this regard are all partial and inadequate. I propose there is a moral reality as well as a physical reality and a mathematical reality underlying the world and the universe, and that human moral life is a search to understand and implement that true nature of morality. I suggest the nature of that moral reality is centred in love, with the idea of *kenosis* ('letting go') playing a key role in the human, moral, and spiritual spheres because of its transformational qualities. This is only one of many intimations of transcendence available to us: these entail qualities in which much more than is necessary is present in the real world in which we live, an abundance leading to wonder and reverence as we realise and appreciate them. An integral view of existence takes these qualities into account. I suggest that true spirituality lies in seeing the integral whole, which includes science and all it discovers, but also includes deep views of ethics, aesthetics, and meaning, seeing them as based in and expressing the power of love. Science can be powerful in the service of this integral view, but must not attempt to supplant it.

Contents

About the author

Abstract

1	The attacks on faith	1
2	Issues of conflict	3
3	Science and reductionism	10
4	The limits of science	21
5	Morality and the source of values	26
6	Intimations of transcendence	39
7	True spirituality	46

1 The attacks on faith

The rise of science over the past 300 years has led to an increasing series of attacks on religious faith, seen by some as a defence of rationality against superstition and irrationality. This has been renewed with vigour in recent times, in particular by Richard Dawkins[1], Daniel Dennett[2], and Viktor Stenger[3]. The swelling of atheist literature is a reaction to a worldwide rise in fundamentalist religion. Doubt about faith and religion has been strengthened by such attacks. What are the intellectual resources and sources of spirituality that can sustain those of faith in these times of uncertainty?

One view[4] is that science has its proper place in dealing with mechanisms—how things work—while religion has its proper place in dealing with completely different issues: meaning, ethics, and metaphysics. Hence there is no possibility of conflict between them, as they deal with quite separate domains. However, this does not seem right: there are at least some places where there are indeed potential or actual conflicts between them. The Dawkins-Dennett-Stenger school claims they do indeed deal with overlapping issues, and there are irreconcilable differences between them when they do so, with science winning all the time. Others[5] have claimed that consonance between science and religion is possible; indeed, they fit together in a complementary way to give an overall view of all reality, with basic agreement in the areas where there are overlaps. This is my view, which I will support in what follows. As I will point out, this means that some of the strong claims of reductionist science (reducing humanity to nothing but a conglomeration of particles and forces) must be wrong; science and rationality are not the answer to all our needs, as some claim.[6] Faith and hope, religious understanding, and spirituality are important aspects of a full humanity.

2 Issues of conflict

Some issues have been problematic for centuries, and remain so. Some used to be areas of conflict, but are no longer so. Others are the site of active conflict, with much debate taking place at present. In this section I will briefly outline what I see as the main issues of each kind. This sets the scene for the later discussion.

2.1 Miracles and prayer

A very longstanding question is how miracles and prayer relate to the regularities of nature. The current science and religion debate adds nothing new to this old theme, and I will not comment on it further. (A lot of the debate hinges on how one regards biblical reports of what happened in the past—an issue in literary understanding rather than the nature of science.)

2.2 A start to the universe?

In the past, one conflict concerned the origins of the universe. There is no reason to question that the universe expanded from a hot Big Bang era at early times. During this expansion from a temperature of about 10^{12} degrees—1 followed by 12 zeros—to the present day, a sequence of physical processes took place that are well understood: nuclear synthesis, the decoupling of matter and radiation, the formation of early stars and galaxies, supernova explosions at the end of the lives of first generation stars, second generation stars, planets and other things, which are pretty much understood.[7] But what is not so clear is what happened before this hot Big Bang epoch. Did the universe have a beginning, or has it lasted for ever? This is still uncertain. It will not be clear till we fully understand quantum gravity—if we ever do. We are certainly not yet there.

It was taken by some that if you could prove the universe had a beginning, this would vindicate biblical claims and so would be good for religion. On the other hand, if you could prove that the universe did not have a beginning, this would be bad for religion—as with Fred Hoyle's theory of the steady-state universe. The current evidence, however, is that the universe did indeed have a start.

Even in the time of St Augustine it was known that this was not a key religious issue, for the fundamental issue is not dependent on whether the universe had a beginning in time. The real question is why the universe exists and has the specific form it does. Why does the universe have this particular form when it could, in principle, have been different? That issue remains a fundamental metaphysical question, irrespective of whether it had a start in time or not. Creation of the universe is not something that happens just at an initial time and then ceases. Keeping the entire universe in existence the way it is, with its very being underpinned by particular laws of behaviour of matter that continue to be valid at all times, is a continuing affair—it is not confined to the start of the universe. It is an ongoing activity all the time the universe is in existence.

One option is the religious one: one can feel quite comfortable with belief in a creative God underlying the existence of the universe, whether it had a beginning in time or not. God could have created the universe in many different ways: it could have been *ex nihilo*, or *ex eternitas*; and the physical way He or She chose to create it is a matter of scientific interest but has no real theological substance. It is the underlying and supporting of existence that matters. The biblical stories are creation stories rather than scientific treatises—important in their metaphysical implications, but not in their scientific content. It is true they are more consonant with a Big Bang view than with the concept of eternal existence, and in that sense the Big Bang may be preferable from the viewpoint of the monotheistic religions; but this is not a logical requisite for the concept of a creator to have validity.

2.3 Darwinian evolution

The more controversial question concerns the origins of life, the mechanism of the evolution of animals and humans. The old religious view—crudely speaking, God sitting at the drawing board designing giraffes and zebras and lions and so on—has gone by the board, and been replaced, as far as all serious biologists are concerned, by our understanding of the Darwinian evolutionary process.[8] Instead, with the modern view of evolution, what you have is the incredible self-creating propensity of nature spontaneously leading to the emergence of complexity and life.

Now this self-creating propensity is based in the laws of physics. In particular, it is based in the way electro-magnetism and quantum theory work. These underlie chemistry, chemistry underlies bio-chemistry, bio-chemistry underlies the way that life comes into being. From the modern viewpoint, if God chose to create humans by the process of designing laws of physics which then make the coming into being of life inevitable, well, that's a wonderful way of doing it. There's nothing wrong with that at all. You can start worrying about the suffering involved in it, but that is part of the bigger problem of suffering in general.

So despite people in the rearguard still fighting out-of-date battles about the issue of evolution[9], there is no fundamental conflict. If a creator shapes laws of physics so that life will come into being, that is an amazing way of getting creation going. There is no serious theological problem.

2.4 Metaphysics of cosmology

But there are still some questions that remain. Firstly, the issue of existence. Why is there a universe? Why are there any laws of physics? Why are the laws of physics the way they are?

What has become clear is that the way life evolves depends on the universe. The universe is a very extraordinary place, in the sense that it appears fine-tuned so that life will exist.[10] A cosmologist can imagine ensembles of universes with all sorts of different properties: bigger or smaller; expanding faster or slower; with different laws of physics, different kinds of particles, different masses of particles; maybe with different laws of physics altogether. As a cosmologist, one can imagine these different universes and think about how they would evolve. And in most of them there will be no life at all: indeed, no process of evolution of any kind will be possible, because there will be no heavy atoms, or no atoms at all; the universe may not last long enough, or may always be too hot; there may be no galaxies at all, and so no stars or planets; and so on.

So what underlies the particular universe in which we exist? Why does it not only exist, but have such a nature that it forms a hospitable habitat for life? This is the issue of the metaphysics of existence, to which I will return later. It is a real issue of contention, which I will revisit. However, it is a rather intellectual concern: it is of importance mainly to those of a philosophical disposition.

2.5 The nature of humanity

The real crunch comes with the issue of being human. What is the essential nature of humanity in the light of modern physics, chemistry, and biology, and, in particular, molecular biology and neuroscience? This is the real potential conflict between science and religion, which is going to go on for a long time.

Here we come up against the views of strong reductionists who produce incredibly thin views of humanity, claiming human behaviour is nothing but the result of our component parts interacting with each other. In the old days it was mainly physicists who stated we are 'nothing but' atoms linked together in complex ways; physical interactions determine all that happens; all higher level interactions (the way we think and live our lives) are mere epiphenomena, consequent on those physics interactions. Nowadays it is much wider: people from sociology, evolutionary theory, psychology and neuroscience are each making claims that they can totally explain human behaviour, and so view humans as being much less than they actually are. They do so with great authority (even though each has different ultimate explanations of human nature). If you disagree with them you are greeted with great derision. In particular, there are philosophers, psychologists and neuroscientists who tell us that consciousness is not real: it is an epiphenomenon. What we think are conscious choices are not real choices. This is a most important area; it is a real threat from the scientific side. I will return to it below.

2.6 The mind and soul

These are more traditional concerns, but are related to the nature of the human mind and the question of consciousness. Is there a soul separate from the body, that lives on somehow after bodily death? These are points of considerable tension, particularly in the light of modern neuroscience, which gives a molecular explanation of how the brain operates. This strongly suggests that what we see is what we get: the mind is based in the mechanisms of brain functioning, and will simply cease to exist when we die. However, despite the enormous amount scientists know about neuroscience and its mechanisms, about the neural correlates of consciousness, the different brain areas involved and so on, we have no idea how to solve the hard problem of consciousness. There is not even a beginning of an approach. So in relation to this and to issues which flow from it, like the question of life after death or reincarnation, one can be agnostic.

Related to this is the issue of religious experiences: are at least some of them real, or are they all self-delusions? Can the mind know entities of a totally other kind through some form of apprehension other than our usual sensory modalities (sight, sound, touch, etc.), or are sensory inputs through the known senses the only way we can obtain information about the external world? Present-day neuroscience strongly suggests the former, but is not completely conclusive; in particular, because the foundations of quantum uncertainty are not yet properly understood. The usual scientists' working hypothesis will be that there is no other way the mind can be in communication with any external entity other than through our usual physically based senses. That opinion will not change unless substantial scientifically credible contrary evidence is given. However, it is possible that if there is such a thing happening, it will by its very nature be inaccessible to scientific probing. It might exist, but be outside what science can test.

2.7 Evolutionary origin of values and religious belief

A key issue is the question of the origin of values. For many decades the social scientists held sway, arguing that values derive primarily from the society in which we live. This is, of course, only a partial answer, for the question then is: so where did society get those values from? More recently there has been a major movement from the evolutionary biology side to claim that evolutionary psychology explains the total origin of values. I will return to this issue later, arguing that this cannot be the total story for a number of reasons; in particular, these arguments may account fairly successfully for the origin of much value-based behaviour, but that is not at all the same as accounting for the origin of normative ethical values themselves. This is, of course, closely related to the age-old problem of evil, as always one of the major perplexities for those who believe in a benign purpose underlying the universe. I will return to this at the end.

Finally, it is claimed that the existence of religion, too, can be explained in evolutionary terms, thereby showing why it exists and hence showing it is not true, as it has been explained away. But this is a non sequitur, and in fact is a specific example of the ***evolutionary origins fallacy***; namely, the belief that once you have an evolutionary explanation of some human behaviour or other, you have completely explained it. This is simply not the case. To see this, realise that this argument applies to any human activity or understanding whatever, *including all scientific theories and evolutionary psychology itself*. That is, if you believe this argument, then (because it is an imperialistic theory that claims to explain everything) there has to be an evolutionary psychology argument explaining the existence and nature of evolutionary psychology too. Does this fact mean that evolutionary psychology is explained away? No, it does not: for the real situation is that an evolutionary psychology explanation for any human activity, theory, or belief whatever is always a partial and incomplete explanation, and its existence is irrelevant to the truth claims of the theory involved. The claim there has to be an evolutionary psychology explanation for the existence of evolutionary psychology does not prove that any specific aspects of that theory are either correct or incorrect! The same holds for an evolutionary psychology explanation of theoretical physics and for religious beliefs.

3 Science and reductionism

Science discovers the physical context of life and the nature of physical causality. It mainly explains things in mechanistic terms. Does this mechanistic view give a complete explanation of what we see, or is there more to life than such an explanation can comprehend? In this section we look at the nature of reductionist scientific explanation, and its completeness in terms of explaining the real world around us. In the next section we look at the limits of scientific explanation overall: what lies in its domain of application and what does not.

3.1 Reductionism and the powers of physics

The basic structure of physical things is well known: quarks make up protons and neutrons, which together form nuclei; these, together with electrons, make up atoms; atoms combine together to make molecules; complex chains of molecules make bio-molecules. If you string these together in the right way you eventually get cells; cells make tissues, tissues make systems, systems make the organism and the organism makes communities. This is the hierarchy of complexity (Figure 1).

Level 8: Sociology/ecology (communities)
Level 7: Psychology (the mind)
Level 6: Physiology (organisms)
Level 5: Cell biology (cells)
Level 4: Biochemistry (biomolecules)
Level 3: Chemistry (molecules)
Level 2: Atomic physics (atoms)
Level 1: Particle physics (quarks)

FIGURE 1 A simplified representation of the hierarchy of complexity (For a detailed survey, see http://www.mth.uct.ac.za/~ellis/cos0.html.)

The common physics view of all this is that bottom-up causation is all there is: electrons attract protons at the bottom level, and this is the basic causal mechanism at work, causing everything else all the way up. In a certain sense that is obviously true. You are able to think because electrons are attracting protons in your neurons. But reductionists tell us this is the *only* kind of causality there is (using the phrase 'nothing but' to emphasise their viewpoint). This is wrong: there are other forms of causality in action in the world; in particular, whole-part causation and human intentionality. Non-reductionist views of science will take them into account, thus taking emergent properties seriously and freeing us from the straightjacket of strong reductionist worldviews.

The important realisation then is that as well as this bottom-up action, there is top-down action in this hierarchy of structure: the top levels influence what happens at the lower levels. They do so *inter alia* by setting the context in which the lower level actions function, thereby organising the way lower level functions integrate together to give higher level functions.[11] An important example is human volition: the fact that when I move my arm, it moves because I have 'told it' to do so. In other words, my brain is able to coordinate the action of many millions of electrons and protons in such a way that it makes the arm move as I desire. Every artifact in the room in which you are sitting, as well as the room itself, was created by human volition—so our minds are causally effective in the world around us. Top-down action from the mind to muscle tissue enables the higher levels of the hierarchy to be causally effective.

It is also important to understand that information is causally effective, even though information is not physical but an abstract entity. Social constructions, too, are causally effective. A classic example of this is the chess set. Imagine some being coming from Mars and watching chess pieces moving. It is a very puzzling situation. Some pieces can only move diagonally and other pieces can only move parallel to the sides. You imagine the Martian turning the board upside down and looking inside the rook, searching for a mechanism causing this behaviour. But it is an abstraction, a social agreement, that is making the chess piece move that way. Such an agreement, reached by social interaction over many hundreds of years, is not the same as any individual's brain state; it exists in an abstract space of social convention, and yet is causally effective. Many other social constructions are equally causally effective, perhaps one of the most important being the value of money. This already is enough to undermine any simplistic materialistic views of the world, because these causal abstractions do not have a place in the simple materialist view of how things function.

Ethics too is causally effective. It constitutes the highest level of goals in the feedback control system underlying our behaviour[12], because it is the choice of which other goals are acceptable. When you have chosen your value system, this governs which goals are inside your acceptable boundary and which are outside. So this abstract entity is causally effective. As a simple example, if your country believes that a death penalty is okay, this will result in the physical realisation of that belief in an electric chair or some equivalent. If you do not believe in the death penalty they will not be there. This lies outside what materialist, reductionist physicists and chemists have in their causal schemes.

The important point is that physics as it currently stands is causally incomplete. It is not able to describe all the causes and effects shaping what happens in the world. For example, physics cannot explain the curve of the glass in my spectacles, because it has been shaped on purpose to fit my individual eyes. The vocabulary of physics has no variable corresponding to the intention that has shaped the spectacles. Because of this, physics cannot explain why the glasses have their particular curvature. This means that physics provides a causally incomplete theory. It cannot describe all the causes acting to shape what happens in the real world.

3.2 Reductionism and the human mind

The human mind and the question of consciousness is one of the most serious potential points of tension between science and religion; and, indeed, between science and the fullness of humanity. There are philosophers, psychologists and neuroscientists who tell us that consciousness is just an epiphenomenon; that we are not really conscious, but are machines driven by unconscious computations, so that what we think are conscious choices are not real. To me, this is the one real threat from the scientific side. Let me quote from Merlin Donald's book, *A mind so rare*: [13]

> Hardliners, led by a vanguard of rather voluble philosophers, believe not merely that consciousness is limited, as experimentalists have been saying for years, but that it plays no significant role in human cognition. They believe that we think, speak, and remember entirely outside its influence. Moreover, the use of the term 'consciousness' is viewed as pernicious because (note the theological undertones) it leads us into error ... They support the downgrading of consciousness to the status of an epiphenomenon ... A secondary byproduct of the brain's activity, a superficial manifestation of mental activity that plays no role in cognition (pp. 29, 36).

> Dennett is actually denying the biological reality of the self. Selves, he says, hence self-consciousness, are cultural inventions ... the initiation and execution of mental activity is always outside conscious control ... Consciousness is an

illusion and we do not exist in any meaningful sense. But, they apologize at great length, this daunting fact Does Not Matter. Life will go on as always, meaningless algorithm after meaningless algorithm, and we can all return to our lives as if Nothing Has Happened. This is rather like telling you your real parents were not the ones you grew to know and love but Jack the Ripper and Elsa, She-Wolf of the SS. But not to worry ... The practical consequences of this deterministic crusade are terrible indeed. There is no sound biological or ideological basis for selfhood, willpower, freedom, or responsibility. The notion of the conscious life as a vacuum leaves us with an idea of the self that is arbitrary, relative, and much worse, totally empty because it is not really a conscious self, at least not in any important way (pp. 31, 45).

But this is not, in fact, what is implied by the science, which has a long way to go before it properly understands the brain, and has made virtually no progress at all in understanding the hard problem of consciousness (however, many of the hardliners even deny there is such a problem). Despite the enormous amount scientists know about neuroscience and its mechanisms, the neural correlates of consciousness, the different brain areas involved and so on, we have no idea of how to solve the hard problem of consciousness. There is not even a beginning of an approach. I find Merlin Donald's writings on these topics convincing. And, personally, I prefer to run this whole argument the other way round, starting with our daily experience.

Consciousness and conscious decisions are obviously real, because that is the primary experience we have in our lives. This is the basis from which all else—including science—proceeds. It is ridiculous to give up that primary experience on the basis of a fundamentalist theory which ignores this fundamental data. And that theory is not even self-consistent, because if Professor Dennett's mind in fact works that way, then you have no reason whatever to believe his theories—for they are then not the result of rational cogitation by a conscious and critical mind. If that were indeed the case, then the entire scientific enterprise would not make sense. Thus I take the causal efficacy of consciousness as a given which underlies our ability to carry out

© The Religious Society of Friends
(Quakers) in Australia, 2008.

ISBN 978-0-9803258-1-2

Produced by Australia Yearly Meeting of the
Religious Society of Friends (Quakers) in
Australia Incorporated

Printed by Uniprint, Hobart.

Copies may be ordered from:
Friends Book Sales, PO Box 181
Glen Osmond SA 5064 Australia.
Email sales@ quakers.org.au

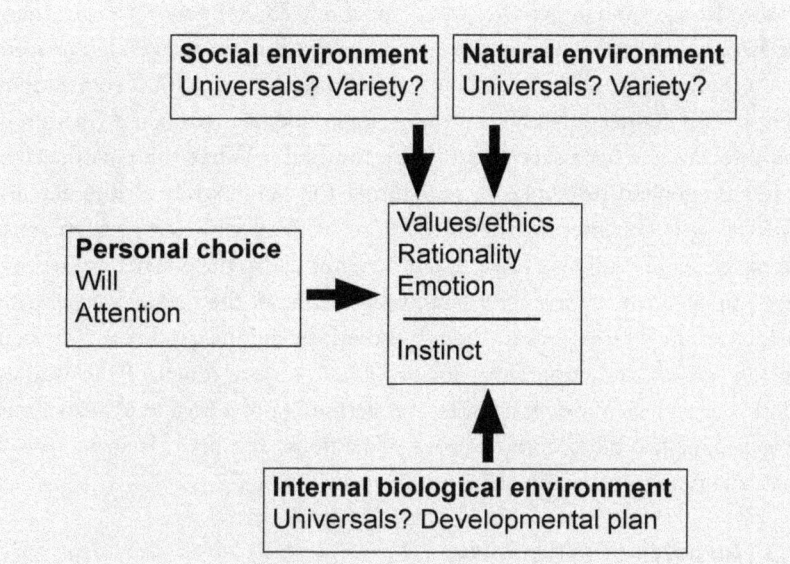

FIGURE 2 Psychological universals are based on universals in the social environment, in the natural environment, and in our inherited biological make-up (underlying a universal human development plan). They all interact with each other to produce the specifics of higher brain functioning via the process of Affective Neuronal Group Selection, shaped by the primary emotions. However, variety in each environment as well as in the genetic inheritance leads to variety of outcomes. Key unresolved issues are: what are the universals in each category, and what variety is there in each of them?

science and to entertain philosophical and metaphysical questions. And, as a consequence, ethical choices and decisions can be real and meaningful.

In terms of human behaviour, we have sociologists and anthropologists who say it's all culture, evolutionary biologists who say it's all genes, others saying it's just physics. But it is not just one of these: it is all of them and more; in particular, our own choices are also causally effective in shaping our own minds. To claim that any of them has no influence would be ludicrous, for they are all involved (Figure 2). What is very interesting is the way that this actually works. The interaction of our brain with the environment, internal and external, shapes the brain. This has to be the case for a simple reason. In the human genome there are about 23,000 genes.[14] From these, we have to construct the entire human body. But there are 10^{11} neurons in the brain, each of which has up to 100, maybe even 1000 connections. That is 10^{14} connections. Hence the genome does not contain a fraction of the information necessary to structure the brain. What the genome does is sets up general principles of structuring the brain, while all the detailed structuring is governed by the interactions we have with our peers, parents, caregivers, environment, and with our own minds. So the genetic influence is very important in setting the basic structure, but all the detailed structuring of each of our brains comes through those interactions. And many of them involve top-down action from society, which shapes much of the way we think[15], as well as our own choices. It is certainly not a case of physics alone, or genes, or the blind computations of neurons. The brain is structured so that true humanity can emerge and function.

3.3 The myth of rationality

Since the time of the Greek philosophers, there has been a perception by some that one could live a purely rational life: that emotion, faith, and hope simply get in the way of rationally desirable decisions.[16] This view was particularly promoted by Descartes, and attained ascendancy with the rise of the natural sciences, with physics taken as a paradigm for the social sciences and rational choice theory an idealised model for human behaviour. It is this viewpoint that underlies much of present-day scientism, views that are taken to deny any spiritual or religious reality.[17] Given such a rationalist view, how can one reasonably have an alignment of religious faith and scientific commitment? How does one hold, without contradiction, a deep and intrinsic respect for evidence and reason, and an equally deep respect for matters of belief?

It is my contention that this view of a purely rational way of existence is a deeply flawed view of how we can conduct both personal and social life. It is not possible to reason things out and make decisions purely on a rational basis. The true situation is much richer than that (see Figure 3).

Firstly, in order to live our lives we need faith and hope[18], because we always have inadequate information for making any real decision. Faith is to

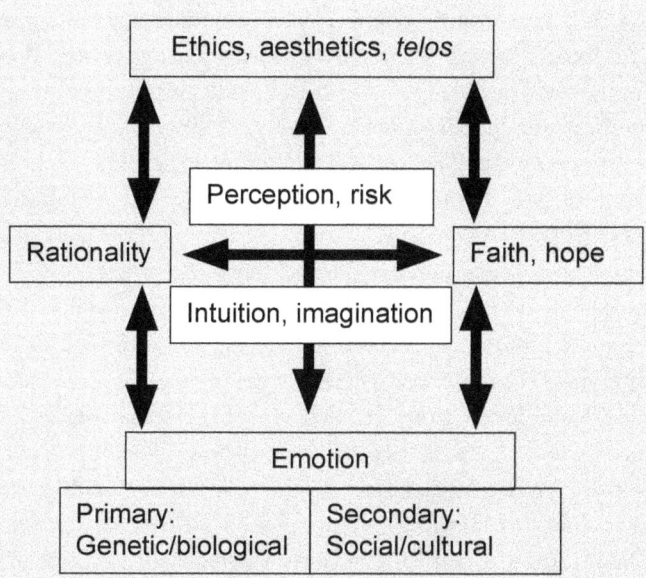

FIGURE 3 Factors affecting actions/decisions: Each of rationality, emotions, ethics, faith and hope are influenced by each of the other, with reason being the key player trying to bring the others into harmony. The instinctive brain (not depicted) underlies this, giving a hard-wired (genetically determined) fast reaction motor channel. Perceptions and attitudes to risk modulate responses. Intuition acts as a short-cut for rationality, embodying an ability to quickly act by activating learnt patterns of understanding in response to recognised patterns; thus intuition is learnt rather than hard wired. The unconscious may feed into this.

do with understanding what is there, hope with the nature of the outcomes. When we make important decisions like whom to marry, whether to take a new job, or whether to move to a new place, we never have enough data to be certain of the situation or the outcome. We can keep gathering evidence as long as we like, but we will never be truly sure as to how many people will buy our product, what the weather will be like, how people will treat us, and so on. Thus our choices in the end have to be concluded on the basis of partial information and are necessarily to a considerable degree based in faith and hope: faith about how things will be, hope and trust that it will work out all right. This is true even in science. When my scientific colleagues set up research projects to look at string theory or particle physics, they do so in the belief that they will be able to obtain useful results when their grant applications have been funded. They do not know for sure that they will succeed in their endeavours. They believe that their colleagues will act honestly. So embedded in the very foundations even of science there is a human structure of hope, and trust.[19]

Cognition and perception are part of the same process. Together with our attitudes to risk, perceptions of how things are now and will be in the future are crucial in making real-world decisions. Do we tend to see things in a threatening or optimistic way? Are we willing to act on the basis of little evidence, or do we demand very detailed analysis before proceeding? This sets the balance we make between rationality on the one hand and faith and hope on the other. Helping us make decisions are intuition[20] and imagination[21]. Intuition is a way of knowing—something to do with understanding and acting. The intuition of a doctor, a motor car mechanic, a football player, a financial analyst, is the deeply embedded result of previous experience and training. It is a fast-track ability to see the guts of the situation long before we have had time to figure it out rationally, embodying in rapid-fire form the results of previous experience and rational understanding. Imagination helps us think of the possibilities to be taken into account in making our rational choices and to envisage what might occur, setting the stage for our analysis of options and choices. But we can never imagine all the options: the completely unexpected often occurs and undermines the best laid plans of mice and men[22], and even the widest lateral thinking only uncovers some of the possibilities.

Secondly, our emotions are a major factor in real decision-making—both the hard-wired primary emotions that are our genetic inheritance from our animal forebears, and the socially determined secondary emotions that are our cultural inheritance from society. As explained so well in Antonio Damasio's writing[23], no decisions are made purely as a result of rational choice; the first factor affecting what we tend to do is the emotional tag attached to each experience, memory, and future plan. For example, the hoped-for joy of successful achievement underlies most work in science; without it, science would not exist. In a full human life, love is one of the most important driving factors, determining how we deploy our rationality. The degree to which one loves another is not a scientifically ascertainable fact.[24] The importance of emotions derives from the fact that the primary emotions have evolved over many millions of years to give us immediate guidance as to what is good for our survival in a hostile environment; they then guide the further development of secondary emotions (telling us what is good for us in terms of fitting into society) and intellect.[25]

Thirdly, we need values to guide our rational decisions; ethics, aesthetics and meaning are crucial to deciding what kind of life we will live. They are the highest level in our goals hierarchy, shaping all the other goal decisions by setting the direction and purpose that underlies them: they define the *telos* (purpose) which guides our life.[26] They do not directly determine what the lower level decisions will be, but set the framework within which choices involving conflicting criteria will be made and guide the kinds of decisions which will be made. Emotional intuitions are absolutely necessary to moral decision-making, but do not fully encompass them—for rational reflection and self-searching are key elements of higher level morality, as one searches for truth and meaning. Indeed, this is all done in the context of overall meaning and purpose (*telos*), for the mind searches all the time for meaning, both in metaphysical terms and in terms of the social life we live. These highest level understandings, and the associated emotions, drive all else.

Our minds act, as it were, as an arbiter between three tendencies guiding our actions: first, what rationality suggests is the best course of action—the cold calculus of more and less, the economically most beneficial choice; second, what emotion sways us to do—the way that feels best, what we would like to do; and third, what our values tell us we ought to do—the ethically best option, the right thing to do. It is our personal responsibility to choose between them (Figure 3), on the basis of our best wisdom and integrity, making the best choice we can between these usually conflicting calls, informed by the limited data available, and in the face of the pressures from society on the one hand (which we must understand as best we can[27]) and from our inherited tendencies on the other. Our ability to choose is a crucial human capacity.[28]

Thus the desire to free us from irrationality leads to the myth of pure rationality, suggesting pure reason alone is the best basis for a worthwhile life. But this is a completely inadequate understanding of causation on which to base a full life. Rationality, faith, hope and doubt as well as imagination, emotions and values are all important in a full understanding of human choices and decisions. They all interact with each other and are causally important in the real world. The key one is values, related to aesthetics and meaning (*telos*): this is what ultimately guides our choices and actions, and so shapes both individual lives and society. Science can help us in the rational part of this cogitation. It cannot by itself be a basis for living a life.

4 The limits of science

Now we consider the limits of science overall: what we learn from it, and what we cannot learn from it.

4.1 The domain of science and its limits

There are many limits to what we can know within the sciences. Knowledge in mathematics is limited by Gödel's incompleteness theorem and by sensitivity to initial conditions (chaos). Knowledge in physics (and cosmology) is limited by observational limits. Knowledge in biology and related sciences is limited by their complexity. Acknowledging this does not deny the power of science: it helps locate the powers of science within its own proper domain.

However, that domain is limited. Many important human endeavours and understandings of necessity remain outside the domain of science; these include the key issues of ethics, aesthetics, metaphysics, and meaning. They transcend scientific views: indeed, they lie beyond the limits of science. Now, when I mention the limits of science, some people immediately say, 'Ah, this is the old "God of the gaps" argument' (Science can't deal with it today, but will be able to in the future: it's just a gap in scientific explanation that will soon be filled in). It is nothing of the sort. It is about boundaries. There are many important concerns for humans which lie outside the boundaries of scientific explanation.

Why are there these boundaries? Because science deals with the generic, the universal, in very restricted circumstances. It works in circumstances so tightly prescribed that effects are repeatable and hence can be reliably duplicated and tested. But most things which are of real value in human life are not repeatable. They are individual events which have meaning for humanity in the course of our history. So science does not encompass either all that is important or, indeed, all that can reasonably be called knowledge.

First, ethics is outside the domain of science because there is no scientific experiment that determines what is right or wrong. There are no units of good and bad, no measurements of so many 'milli-Hitlers' for an action. Correspondingly, aesthetics is outside the boundaries of science. No scientific experiment can say that something is beautiful or ugly. Both are related to the way we understand meaning in our lives—what is valuable and what is not, what is worth doing and what in fact makes life meaningful. These are areas of life which cannot be encompassed in science: they are the proper domain of philosophy, religion, and spirituality. And meaning relates to metaphysics: whatever it is that underlies the nature of existence, and in particular whether there is some kind of transcendental reality underlying the physical world. We deal with each of these issues in what follows.

4.2 Metaphysical issues

What underlies existence? What underlies the creation of the universe and the specific nature it has? This metaphysical issue was mentioned above, and I now focus on the 'anthropic' question, already mentioned there: the universe appears to be fine-tuned so that life can exist. Why should this be so?

Life as we know it on Earth would not be possible if there were very small changes indeed to either the nature of physics, or to the universe itself. There are all sorts of fine-tunings in physics that must be obeyed if the emergence of complexity is to be possible. For instance, both the difference in mass between the proton and the neutron and the ratio of the electro-magnetic to the strong nuclear force have to lie in a very narrow range if atoms are to exist; and without heavy atoms, no normal life can come into being. If you tinker with physics, you may not get anything heavier than hydrogen; or maybe if the initial conditions of the universe are wrong it won't create suitable habitats for life. For example, the universe is not slowing down as we expected, but is expanding faster and faster due to a cosmic force known as the 'cosmological constant' or 'quintessence'. We do not know why this force is there, but we do know that if it were substantially bigger than it is, there would be no galaxies at all, no planets, no life. There are many coincidences which, taken together, make the universe a suitable habitat for life. It could have been different—and then we would not exist. There are all sorts of things that can go wrong if you are the creator trying to create a universe in which life exists.[29]

Some scientists do not see this as a valid issue, regarding it as unscientific and therefore not worth considering. But for those concerned with ultimate issues of existence and causation, it is a serious metaphysical question. (It cannot be answered by science per se, because no scientific experiment can determine why the universe is as it is.) There are essentially four ways of trying to answer it.

One is ***pure chance***. Just by chance everything worked out right. That is a logically tenable position, if you like to live with extremely thin philosophies. But it has no explanatory power; it doesn't get you anywhere. So it is not an argument that is popular in scientific or any other circles.

The second option is ***necessity***. Although it looks as if things could have been different, this is not the case; there is, in fact, only one possible form of physical reality, namely that we see around us. If we understood physics deeply enough, we would be able to prove that no other physical system is self-consistent. This is an attractive option which physicists would love to make real, but they have not succeeded; on the contrary, modern physics seems to envisage more and more possibilities, most of which simply do not allow life to exist.[30] The hope that one could prove only one unique form of physics is possible seems to have failed. And if it were to succeed, the anthropic issue would return with a vengeance: why should the unique possible physical state that can exist be one that allows life? That would remain completely unexplained, as a most mysterious fact about the nature of physics.

The third option is the idea of a *multiverse*. The proposal is that this is not the only universe; rather, there are millions and millions or even an infinite number of universes, all with differing properties (or perhaps there is one huge universe with different domains looking like our expanding universe but with different constants, different rates of expansion and so on). So although there is an incredibly small chance of a universe existing that will allow life, if there are enough universes, or at least expanding universe domains with different properties, life becomes essentially inevitable.[31]

This is popular nowadays as the only scientifically based approach that seems to explain the anthropic issue. It obtains support from the idea of a chaotic form of inflation in the very early universe[32], which is said to follow from well-known physics. However, in fact, that supposed physics is untested and simply has not been proven to be correct. The problem with this proposal overall is that these other universes cannot be observed. They are beyond the part of the universe that we can see or detect by any means whatever, so whatever is said about them can never be proven wrong. That makes this a metaphysical rather than a scientific solution.[33] No one can actually *prove* any single other universe exists, let alone many millions (or an infinite number, as is often claimed); nor can one determine what the properties of all these other universes are (if, indeed, they do exist). The distinguishing feature of science is that you can test its proposals, and there is no way of testing this proposal. So belief in the existence of a multiverse is in fact an exercise in faith rather than science.[34] Furthermore, if it could be shown to be true, this would not solve the ultimate metaphysical issue, which will simply recur: Why this multiverse rather than that one? Why a multiverse that admits life rather than one that does not?

The final option is the good old ***designer*** argument: the way the universe functions reveals intention, the work of some kind of transcendent power or force. Life exists because this fine-tuning took place intentionally.[35] This is the position of all the main monotheistic religions.

The key point to make here is that these are all logically possible and acceptable options, and no scientific test can disprove any of them, or prove which are correct explanations. (Note that they are not necessarily incompatible: several might apply at the same time; for example, a designer could choose to create a multiverse rather than a single universe.)

Which is the case cannot be scientifically determined; our belief about it is a matter of faith. Science per se cannot resolve the issue. However, the further data considered below can be taken to support this view: experience from our aesthetic and moral and spiritual lives can help us choose between these options.

5 Morality and the source of values

The source of values is a key point. As pointed out above, it is not possible to reason things out only on a rational basis because, apart from anything else, there are value choices that come in and guide the decisions made. Rationality can help when we have made these value choices, but the choices themselves, the ethical system, must come from outside science.

5.1 The origin of values

The various scientific proposals in this regard are all partial and inadequate. Science cannot provide values, for the simple reason that there is not any scientific experiment that relates to right and wrong, to good and bad. These are outside the domain of scientific experimentation.

Two things are crucial here. Firstly, values are not the same as emotions; what we feel like doing at some instant may or may not be what is ethically right, 'road rage' being a classic example. Some evolutionary psychologists seem to tend to confuse these issues, assuming values are subsumed under emotions; but this is not the case. They are crucially different. Ethical values have a normative dimension that cannot be present in emotions per se (although emotions are one of the factors helping us understand normative values).

Secondly, guiding values cannot be arrived at purely rationally. They are decided on the basis of an interlocking set of factors that include emotions and rationality, but also a broad-based understanding of meaning based in our total life experience, which is surely data about the way things are. Humans have a great yearning for meaning[36], and ethics embodies those meanings and guides our actions in accordance with them; but ethics, aesthetics, metaphysics, and meaning are outside the competence of science because there is no scientific experiment which can determine any of them. Science can help illuminate some of their aspects, but is fundamentally unable to touch their core. They are the proper domain of philosophy, of religion, and of art, but not of science.

There is a great deal of confusion about this, particularly in the case of ethics. Socio-biology and evolutionary psychology produce arguments which claim to explain where our ethical views come from. There are many problems with those attempts, the first being they do not explain ethics, they explain it away, another being that these arguments ignore key social effects and culture, as well as the role of our individual religious and moral experiences. If the true origin of our ethical beliefs lay in evolutionary biology, ethics would be completely undermined, because you would no longer believe that you had to follow its precepts; you could choose to buck the evolutionary trend. Furthermore, if you did follow those precepts you tend to rapidly end up in dangerous territory, the domain of eugenics and social Darwinism. That has been one of the most evil movements in the history of humanity[37] (a fact that Darwinian propagandists conveniently ignore, when extolling the evils of religion).

In recent times, the possibility of evolutionary psychology explaining altruism via kin selection has been a major theme.[38] There are two problems with this as a proposal for the origin of genuinely altruistic ethics. Firstly, if altruism extends only to kin and those whose genes will be preserved by acts of sacrifice, then by definition it excludes all outgroups—and hence cannot by its very nature explain the kind of ethic that says 'Love your enemy'.[39] It implies hostility to those with competing genes—hence by its nature providing a basis for enmity and hostility to outsiders, and ultimately for war against others.

Secondly, the concept of altruism invoked by the evolutionary psychologists is a pale cousin of the true thing as envisaged in religion, where altruism by its very nature is conceived of as having no reward.[40, 41] This means that it is not possible to explain its origin by evolutionary psychology, because the key causal link in terms of promoting specific genes—based on a reward mechanism that tends to preserve those genes—is missing, by the very definition of the nature of deep altruism. This kind of argument can only explain shallow altruism.

Actually, this argument is just another example of the evolutionary fallacy (see section 2.7 above). And in the end, challenging evolutionary biologists on this issue is simple. If a scientist claims that science can provide a basis for ethics[42], say to them, 'Tell me, what does science say should be done about Iraq today?' You will get a deafening silence, because science cannot handle ethical questions. Answering this important question lies outside the domain of science per se.

5.2 Moral realism

So where do moral values arise? I propose there is a moral reality as well as a physical reality and a mathematical reality underlying the world and the universe.[43] There is a standard of morality which exists (in 'reality', just waiting to be discovered) which is valid in all times and places, and human moral life is a search to understand and implement that true nature of morality.[44]

Thus I take the position of moral realism, which argues that we do not invent ethics, but discover it. Whole sociological schools suggest that we invent ethics—it is socially determined. That route ends up in total relativism, where it is impossible to say that any act is evil. All you can say then is that some people were differently brought up; you cannot judge any acts as being good or evil.

If you believe truly that some acts are good and some (those of Hitler, for example) are bad, in a sense that is true in all times and places and that can be validated across cultures, then you have an indisputable ability to distinguish good from bad, and that is a statement of moral realism. And the intriguing thing is that even some of the arch-enemies of religion believe in a moral reality. Richard Dawkins does, because he states categorically that religion is the cause of much evil[45]; and he can only say that if he has impartial standards of good and evil to use in making that judgment. This view is supported by Viktor Stenger[46], who states that science can prove evil exists.

This is a category error—science cannot do any such thing—but, nevertheless, it shows clearly he too believes in trans-cultural universal standards of good and evil. This means that we do not invent ethics; rather, we discover it (much as the same may be claimed for mathematics). It is already in some sense existent out there.[47]

Now it is true that in human history, socially accepted standards of morality are continually changing and 'exist' by virtue of social agreement—for instance, societies seem to have been quite happy (and maybe some still are) to condone genocide, slavery, etc., while still claiming to be virtuous. The viewpoint here is that these socially accepted standards are not the true morality, but are our best human understandings of it at a particular time and place. Much of human history can be read as our growing understanding of the true unchanging deep morality as human beings became more and more conscious of other cultures and worldviews. Of course there have been backslidings from time to time: but, nevertheless, a major forward thrust in ethical understanding can be seen in recent human history.[48]

5.3 Deep morality

If there is a moral reality, then what is the nature of the moral ethic implied? I suggest the nature of that moral reality is centred in love, with **kenosis** ('self-emptying'—a letting go of the self and selfish interest) playing a key role in the human, moral, and spiritual spheres because of its transformational qualities.[49] This is quite different from the shallow ethics on which everybody agrees and which socio-biology can largely explain. *Kenosis* is here understood not just as letting go or giving up, although this is a key element, but as being prepared to do so in a creative way for a positive purpose in tune with a creative and loving worldview. Thus it is seen as *a joyous, kind attitude that is willing to give up selfish desires and to make sacrifices on behalf of others for the common good, doing this voluntarily in a generous and creative way, avoiding* *the pitfall of pride, and guided and inspired by love.* It is based on a realisation of the preciousness of each human being.[50]

When experienced and understood, this kind of quality with its transforming nature is self-validating; it has an intrinsic rightness and deepness that cries out to be recognised. It is far deeper than any of the other views of morality.

Neither a strictly science worldview, nor an impersonal world that is in the end based only in particles and their interactions, can provide a basis for this kind of nature of being. It has to have a deeper nature—to be somehow inbuilt into the nature of reality in some deep way, rather than arising by sheer chance as a happenstance through impersonal interactions of particles and forces. Now the most obvious basis for such a worldview is that it expresses the underlying deep nature of reality—which, in order to express such values and meanings, must of necessity involve the kind of values that are embodied in the deeply spiritual views of religion.

Indeed, this kind of view is, for example, the core of true Christianity: the suffering of Christ on the Cross is a *kenotic*, self-sacrificial giving up on behalf of the other. This *kenotic* understanding became explicit in the life of Jesus, uncertain as to his destiny and mode of operation, as he examined his possible choices of action in the desert and saw the transcending possibilities opened by the way of love. Considering in the desert the options of using the power with which he was endowed to satisfy the creature wants of himself and his human brethren, or winning the kingdoms of the world by establishing an earthly monarchy, or providing irresistible evidence of his divine mission so that doubt would be impossible, Jesus came to the key insight described by William Temple as follows[51]:

> Every one of these conceptions contained truth. Yet if any or all of these are taken as fully representative of the Kingdom, they have one fatal defect. They all represent ways of securing the outwards obedience of men apart from inner loyalty; they are ways of controlling conduct, but not ways of winning hearts and wills. He might bribe men by promise of good things; he might coerce men to obey by threat of penalty; he might offer irresistible proof; [but] all

these rejected methods are essentially appeals to self-interest; and the kingdom of God, who is love, cannot be established that way ... The new conception which takes the place of those rejected is that the Son of Man must suffer. For the manifestation of love, by which it wins its response, is sacrifice. The principle of sacrifice is that we choose to do or suffer what apart from our love we should not choose to do or suffer ... The progress of the Kingdom consists in the uprising within the hearts of men of a love and trust which answer to the Love which shines from the Cross and is, for this world, the Glory of God.

This is the paradoxical way of true transformation, described by Parker Palmer as follows[52]:

On the cross our small self dies so that true self, the God self, can emerge. On the cross we give up the fantasy that we are in control, and the death of this fantasy is central to acceptance. The cross above all is a place of powerlessness. Here is the death of the ego, the death of the self that insists on being in charge, the self that is continually attempting to impose its own limited version of order and righteousness on the world ... this is the great mystery at the heart of the Christian faith, at the heart of the person of Jesus, of Gandhi, of Martin Luther King Jr: the power of powerlessness ... Emptiness is a key word describing the appearance of acceptance ... Jesus on the Cross emptied himself so that God could enter in.

I believe that each of the major world religions has a spiritual tradition that believes seriously and deeply in a *kenotic* ethic.[53] When I spoke about this once in California, a gentleman came up to me in great excitement and said, 'That was a terrific talk, you spoke like a true Muslim'. I was amazed. He was the director of the Muslim Study Centre in London. I heard a talk given by the Chief Rabbi of Great Britain in which the same spirit was expressed, and I told him, 'You talk like a Quaker'. He said, 'I choose to take that as a compliment'. The same understanding is deeply embedded in the Hindu tradition in which Mahatma Gandhi grew up. There is good reason then to see this concept as of universal validity, a deep aspect of reality true at all times and places, and hence as a key to understanding the deep nature of creation.

5.4 *Kenosis* and ethics

I believe this view is deeply embedded, in particular, in the Quaker attitude towards war and peace:

> The Quaker testimony concerning war does not set up as its standard of value the attainment of individual or national safety, neither is it based primarily on the iniquity of taking human life, profoundly important as that aspect of the question is. It is based ultimately on the conception of 'that of God in every man' to which the Christian in the presence of evil is called on to make appeal, following out a line of thought and conduct which, involving suffering as it may do, is, in the long run, the most likely to reach to the inward witness and so change the evil mind into the right mind. This result is not achieved by war.[54]

This is the most fundamental way to fight evil, with its purpose: the transformation of evil intentions to good, and the redeeming of those who do evil into what God intended them to be. This is not achieved by military force or by buying people, it is not even achieved by intellectual argument or persuasion. It is achieved by touching them as humans. This is achieved particularly by sacrifice on behalf of others, as exemplified in the life and work of Martin Luther King, Mahatma Gandhi, and Desmond Tutu.

The attitude of deep ethics is not that you are always self-sacrificing on behalf of others, it is that you are prepared to do so if and when it will make a strategic difference. That is a significantly different thing. There are times when it is the only thing which will make a difference.

One characterisation of the way demonstrated is that it always transcends the immediate problem by changing to a context of self-giving loving, thus moving to a higher plane where love and forgiveness are the basic elements. This change of perspective and context has the possibility of transforming the situation. This does not mean compromising truth; it does mean creating hope of reconciliation, in that all activities can be forgiven, so that anybody can be redeemed. Our acts and spirit of forgiveness should demonstrate this, if necessary through loving sacrifice. It means being willing to love the enemy rather than giving in to hate, which has the power to transform us into a hateful kind of person. Thus it is a refusal to give in to the hatred embodied in the enemy image. This may mean dealing with those who commit unspeakable evil. An example is that the Quaker Rufus Jones went to try to visit Hitler with two other Quaker leaders because he believed he could convince the Nazis to allow Jews to leave Germany. Jones got as far as speaking with Reinhard Heydrich, a top Gestapo leader, and he did have some success in promoting this idea (although subsequent events nullified their small project).[55]

And here is the hardest part. It is easy to see that respecting the light of God in a person enables one to help and support those oppressed. But the point is that this theme applies to the oppressors too. They too are human, they too have the light of God in them. If in our pursuit of the rights of one group, we turn in fury on their oppressors and kill them or torture them because of what they have done, then we too have fallen into the fatal trap: the infection of hate will have taken hold of us too, and made us behave as the oppressors did. True respect for every person does not excuse or condone evil, but also does not deny the humanity and spark of vital life and the possibility for change in even those persons who are carrying out the foulest deeds. That is the real test and the real foundation; it is the basis of that transforming spirit which is the basis of social and political miracles. The attempt to follow this way is incredibly difficult. It will be much easier if we are able to practise the presence of God, and particularly an awareness of the Light of Christ within every person. Indeed, the two are inexorably linked, for if we are aware of that Presence and its loving nature, we will see the present problems in this profound context, and this will transform its nature for us.

Forgiveness is a huge step on the way. It is not the whole way but it is part of it. This involves the ability truly to see others as fully human instead of seeing them through the enemy image which allows you to treat them as sub-human. This is the subject of a book I wrote with theologian Nancey Murphy, called *On the moral nature of the universe*.[56] It has on its cover a picture of Dresden in 1945. Many of you will know that picture: the burnt-out ruins of the town below and in the foreground a stone angel, with hands stretched out in a gesture of pity over the shattered remnants of Dresden. We used that picture to explain the alternative to a *kenotic* view of life. If you implacably see the other as enemy, it is in those ruins that we all will end.

As with all pacifist literature, many have suggested that our argument is impractical in the real world. In response to that, I will read a most remarkable document by David Christy, who got in touch with me after I was awarded the Templeton Prize:

> In 1967 I was a young officer in a Scottish battalion engaged in peacekeeping duties in Aden, which is now in Yemen. The situation was similar to Iraq with people being killed every day. As always those who suffered the most were the innocent local people. Not only were we tough but we had the firepower to pretty well destroy the whole town had we wished, but we had a commanding officer who understood how to make peace and he led us to do something very unusual, not to react when we were attacked. Only if we were one hundred per cent certain that a particular person had thrown a grenade or fired a shot at us were we allowed to fire. During our tour of duty we had 102 grenades thrown at us and in response the entire battalion fired the grand total of two shots, killing one grenade thrower. The cost to us was over 100 of our own men wounded, and, surely by the grace of God, only one killed.
>
> When they threw rocks at us we stood fast, when they threw grenades we hit the deck and after the explosion we got to our feet and stood fast. We did not react in anger or indiscriminately. This was not the anticipated reaction. Slowly, very slowly the local people began to trust us and made it clear to the 'local terrorists' that they were not welcome in their area.
>
> At one stage neighbouring battalions were having a torrid time with attacks. We were playing soccer with the locals. We had in fact brought peace to our area at the cost of our own blood. How had this been achieved? Principally because we were led by a man who every soldier in the battalion knew would

die for him if required. Each soldier in turn came to be prepared to sacrifice himself for such a man.

Many people may sneer that we were merely obeying orders but this was not the case. Our commanding officer was more highly regarded by his soldiers than the general. One might almost say loved. So gradually the heart of the peacemaker began to grow in each man, in a determination to succeed whatever the cost. Probably most of the soldiers like myself only realised years afterwards what had been achieved.

Changing enemies into friends is the basis of true security. All the time, even if we have to resist them with force, we can be offering the other person a way out, offering them their full humanity, rather than saying, 'You are irredeemable'. I think that is the key. All the time it must be clear to that person: 'I'm going to do everything I have to do to stop you, but I'm not treating you as sub-human'. There must always be that chance, that opportunity for change and reconciliation.

5.5 The wider nature of *kenosis*

Overall, I see *kenosis* as a generic principle with much wider implications in life overall than just in the ethical sphere discussed above, where it is indeed the basis of transformation. It entails learning to give up that which we desire to cling to, accepting the implied loss as the basis of greater good; this leads to a profound view of how to live at all levels of life.

Community life: Self-emptying is the basis of family life, in particular the willingness of the mother to give up her life on behalf of her child. But even more important is her willingness to let go of the child once it has grown up—perhaps a more difficult task. *Kenosis* is the basis of community in general, for that involves giving up one's own needs to some degree on behalf of the welfare of those around. This is important in personal life as a component of friendship (being together, but allowing the other space), in political and cultural life—making space for others, acknowledging the dignity of difference.[57] It is a key aspect of sustainable economics, where it enables moving from destructive self-centredness to a community sharing that transcends the economy of scarcity.

Learning is based on *kenosis* in the sense that you have to give up your preconceptions about the way things are in order to see things as they really are. The person who knows all the answers is unable to see what is in front of them, and cannot learn. Learning is based in letting go of previous vision.

Art: Self-emptying is crucial in the artistic endeavours of people who are truly creative. As they work on plays, books, sculpture, whatever, they start by shaping the thing in front of them, but at a certain stage it takes on its own personality and integrity. Then the artist's need ultimately is to respond to it, to respect its integrity, and not to impose the self on it. The key step is letting go of your own vision in order to let the nature of the creation come to fruition:

> Instead of rules, the great artist follows the spirit, the internal flow, the nature of the thing at hand ... True action, effective action, action that is full of grace, beauty, and results, is action based on discernment of and respect for the nature of the other.[58]

Religious life: One must be prepared here to question one's faith and to let go of it, to see what then comes back and remains. According to Robert Bellah[59]:

> The deepest truth I have discovered is that if one accepts the loss, if one gives up clinging to what is irretrievably gone, then the nothing which is left is not barren but is enormously fruitful. Everything that one has lost comes flooding back out of the darkness, and one's relation to it is new—free and unclinging. But the richness of the nothing contains far more, it is the all-possible, it is the spring of freedom.

To take a further, very practical example from the Quaker Meeting, consider the question: should you speak in Meeting? If you have an urging to speak, a voice emerging from inner depths, I suggest that the thing to do is to consciously give up the need to speak, to let go of the need to be heard, and then to listen in the silence, to wait, to hear. Then you can see if the need to speak still remains after you have given it up, and if it does, you can then respond to it.

Discernment: This is an example of the question of discernment. The problem is that throughout history many people have felt strongly that they were being led by God, but you can tell by their actions that in some cases they were being led not by the loving God of Jesus, but rather by some other God or by their own self-centredness. Many evils have resulted; inter alia, the Crusades and the Inquisition were carried out in the name of God, and the policy of *apartheid* was supported by a group of churches. The crucial issue is discernment, the testing of such urgings to see if they are really the true voice. And that is done by giving it up, and then looking at it without pressure to try to discern the real nature of what should be done.

There is a clear link with science here, because the strength of science comes through its process of testing to see if its conclusions are true, which involves the readiness to discard theories (no matter how dearly held) that do not conform to experiment. The real challenge for us is to test our spiritual leadings for their veracity in a similar way. Quakers have various mechanisms which we have developed precisely for this purpose. One of the reasons we have to listen to others even when we disagree with them is the belief that, no matter how irritating they are, they may have something to which we should be listening. They may have seen the Light in a way that we have missed.

Overall, in each case *kenosis* has a transforming nature, lifting the situation to a higher plane, thereby opening up possibilities that were not there before. Parker Palmer summarises it as follows[60]:

> The cross signifies that the pain stops here. The way of the cross is the way of absorbing pain, not passing it on; a way which transforms pain from a destructive impulse to a creative power. When Jesus accepted the cross, his death became a channel for the redeeming power of love ... the way of the cross means letting the pain carve one's life into a channel through which the healing spirit can flow into the world in need ... the great paradox of the

crucifixion is Christ's victory over the illusion that death is supreme ... beyond illusion lies a fuller truth which can be glimpsed only as our falsehoods die. Only as we have the faith to live fully in the midst of these painful contradictions will we experience resurrection and the transformation of our lives.

6 Intimations of transcendence

6.1 Awareness of transcendence

The experience of *kenosis* seems to point beyond the everyday world to a set of higher processes that underlie the way reality works. This kind of experience can, in my view, be regarded as one of many *intimations of transcendence* available to us—perceptions of a deeper kind of existence lying behind the surface appearance, giving a grounding for meaning, morality, and purpose (see Figure 4). [61]

I think everyday experience gives such hints in many ways. In some sense—not in a scientifically provable way, but as an intuitive kind of feeling—the beauty and glory of what exists is more than is necessary. There is an over-abundance leading to wonder and reverence as we realise and appreciate it.

It did not have to be that way. In Cape Town one day I was watching the waves coming into Clifton Bay, and there were fourteen dolphins surfing there. One of them was doing back somersaults as he surfed. They were simply having fun! There was no serious purpose in it, it wasn't going to help them survive or get them food. It was just for the joy of it. It was an

Aesthetics The beauty of transcendence	**Love and joy** Father/mother Companion/friend

Spiritual awareness Awe, wonder at creation Transcendent reality experienced as immanence

Creativity Co-creators, inspiration, humour, ability	**Ethics** Right and wrong, justice and forgiveness

FIGURE 4 Intimations of transcendence: The various dimensions of the intimations of transcendence available to us. These are patterns of understanding that are more than the necessary minimum.

unnecessary expenditure of energy—and it was joyous and wonderful. This is an example of how life is often much more than the necessary minimum. The complex web of interactions and values is much richer than any reductionist argument tries to persuade us is the case. These entail qualities in which much more than is necessary is present in the world in which we live.

Then, again, there is the obvious that we take for granted and so do not see. The nature of our existence and, indeed, the very fact of our existence is one of the most wondrous things one can imagine, and, indeed, the single most important fact about our lives; yet we forget this and take it for granted. Here is a Buddhist meditation that tries to help us remember this[62]:

> I like to walk alone on country paths, rice plants and wild grasses on both sides, putting each foot down on the earth in mindfulness, knowing that I walk on the wondrous earth. In such moments, existence is a miraculous and mysterious reality. People usually consider walking on water or in thin air a miracle. But I think the real miracle is not to walk on water or in thin air,

but on earth ... Every day we are engaged in a miracle which we don't even recognise: a blue sky, white clouds, green leaves, the black, curious eyes of a child—our own two eyes. All is miracle.

And here is a beautiful meditation on this theme by Antoine St Exupery[63]:

I say to myself as I watch the niece, who is very beautiful: in her this bread is transmuted into melancholy grace. Into modesty, into a gentleness without words ... Sensing my gaze, she raised her eyes towards mine, and seemed to smile ... A mere breath on the delicate face of the waters, but an affecting vision. I sense the mysterious presence of the soul that is unique to this place. It fills me with peace, and my mind with the words: 'This is the peace of silent realms'. I have seen the shining light that is born of the wheat.

This existence is based on the amazing physical nature of reality as explored by science: physics, chemistry, cosmology all combine to make this possible. The nature of the world and the universe is beautifully crafted to allow us to exist and, indeed, to make our existence virtually inevitable. We mostly take this for granted; but this is what has led to our existence; we owe everything to it.

6.2 Beauty and transcendence

Beauty is for many a way of experiencing a transcendent reality, which is perhaps why people value it so much and some devote their entire lives to it. This is expressed by Rufus Jones in this way[64]:

Perhaps more wonderful still is the way in which beauty breaks through. It breaks through not only at a few highly organised points, it breaks through almost everywhere. Even the minutest things reveal it as well as do the sublimest things, like the stars. Whatever one sees through the microscope, a bit of mould for example, is charged with beauty. Everything from a dewdrop to Mount Shasta is the bearer of beauty. And yet beauty has no function, no utility. Its value is intrinsic, not extrinsic. It is its own excuse for being. It greases no wheels, it bakes no puddings. It is a gift of sheer grace, a gratuitous largesse. It must imply behind things a Spirit that enjoys beauty for its own sake and that floods the world everywhere with it. Wherever it can break through, it does break through, and our joy in it shows that we are in some sense kindred to the giver and receiver.

I believe that for many the experience of great beauty is an immediate, striking way of experiencing transcendence. It is expressed, for example, in the book *Timeless beauty* by John Lane.[65]

6.3 Spiritual experience

For many, this leads on to what many people believe are genuinely spiritual experiences.[66] These are of many kinds for different people, for our many different cultures and viewpoints lead to many different interpretations of the sacred[67]; the various religious faiths have many ways of approaching the nature of spirituality and of experiencing it.[68] The proposal here is that at least some of those experiences—the gathered Meeting for Worship, for example[69]—are what they seem to be: genuine intimations of transcendence. This does not mean all are—discernment is needed to see which are genuine and which not. But it opens the possibility that this kind of experience (of which there is a huge amount) is genuine.

For those of Christian belief, the life of Jesus as a historical event (i.e. the incarnation) is the most profound locus where the transcendent realm breaks into our space-time. A passage that conveys this idea strongly to me is the start of St John's Gospel:[70]

1. In the beginning was the Word, and the Word was with God, and the Word was God.
2. He was in the beginning with God;
3. All things were made through him, and without him was not anything made that was made.
4. In him was life, and the life was the light of men.
5. The light shines in the darkness and the darkness has not overcome it.

9 The true light that enlightens every man was coming into the world.
10 He was in the world, and the world was made through him, yet the world knew him not.
11 He came to his own home, and his own people received him not.
12 But to all who received him, who believed in his name, he gave power to become children of God;
13 Who were born, not of the will of the flesh nor of the will of man, but of God.
14 And the word became flesh and dwelt among us, full of grace and truth; we have beheld his glory, glory as of the only Son from the Father.

This kind of view was deeply embedded in the life of George Fox and the early Quakers. It is still the view of many in the Quaker movement.

6.4 The nature of evidence

Why should we believe any of this has anything to do with reality? The claim I make is that the kinds of personal world experience we each have are certainly data on the nature of reality, because we live in and, indeed, are part of reality. They do not have the quality of the strictly repeatable experiments that science engages in—they are much richer than that. And they can have the transcendent quality I have outlined above. The fact that such a quality can exist is itself a statement about the nature of reality.

For many, a deeply religious worldview is crucial in understanding our lives and setting values, this worldview being based in our personal life experience, including our experience of a faith tradition and community, religious texts, and inspiring leaders. All of these are data that help us understand our situation and our lives. Something of this kind is essential to our wellbeing and proper fulfilment, because ethics and meaning are deeply intertwined. This does mean taking a stand as regards the different interpretations of the various religious traditions; for example, relating to the generous and loving nature of the heart of true Christianity[71], rather than the oppressive nature of some of its manifestations. That self-emptying vision embodied in the life of the great religious and spiritual leaders of all faiths[72] provides an inspiring basis for a deep ethics and for life. It can provide a deeply meaningful vision of the nature of reality. This view cannot be *proved* to be true—but it is supported by much experience that has considerable persuasive power as a whole.[73]

6.5 The problem of evil and suffering

But what about the counter evidence? The problem of evil is a key issue here—one of the oldest facing religion. This is not the place for any real attempt at a discussion, except to say that: first, we need to remember that there is great good in the world as well as evil; and, second, if God plans to create conditions allowing people to come into existence who can exercise free will and are able to use that free will to love others and to love God, then that choice constrains what is possible in other ways—in particular, you cannot offer independent beings free will and also prevent them from doing evil. Similarly, you cannot create physics and biology that will lead to the existence of humanity that can exercise free will, and also have a world without pain and death. One can't have these possibilities without also having the possibility of evil. This is one of the ultimate paradoxes.

But still, what about all the ugliness and suffering? One can even wonder if on average the beauty and love outweigh the suffering. What makes one optimistic? Surely theodicy is the big unanswered theological question, and one can doubt that it's answered by beautiful sunsets or even cute grandkids, with their marvellous quality of vital life. On what shall hope be based?

I suppose the only answer for the Christian is in the life of Jesus. In some sense, the answer to evil is the image we have been given of God suffering voluntarily on our behalf—freely accepting that suffering in order to create a greater good. That act not only shows the way for us to go, but also shows that God himself follows that way and accepts the suffering, thereby transcending death. In the end, it is the contemplation of the Cross that is the solution to evil—not in an intellectual sense, but in the sense of allowing us to share both the pain and the glory with God. That sharing is the true nature of *kenotic* love. If one believes in a loving God, then one accepts that somehow it will indeed come out right—and that is the promise given us of resurrection, whatever that may mean.[74] For those of other faiths, all one can do is have faith in the goodness of what underlies creation, and hope in some kind of positive ultimate resolution.

7 True spirituality

7.1 Embracing the whole

I suggest that true spirituality lies in seeing the integral whole, which includes science and all it discovers, but also includes deep views of ethics, aesthetics, and meaning, seeing them as based in and expressing the power of love.

The Methodist theologian Walter Wink[75] presents a brief discussion of different worldviews in his book, *The powers that be*. He says that 'we may embrace the integral worldview as a way of reuniting science and religion, spirit and matter, inner and outer'. Science can be powerful in the service of this integral view, but must not attempt to supplant it. We always need to remember that there are limits to what we can know about both science and religion.

Both science and broader aspects of human experience are important to being a fully rounded human being. We need to incorporate both of them in our understanding of life and existence. Even if you are not a scientist, it is worth trying to find out about science because it tells us so much. But this does not mean having to deny religion or indeed humanity. The religious life adds an enormously important dimension to humanity, individually and collectively, when approached in a non-fundamentalist way.

Our broader experience can give us a relation to spiritual issues with many dimensions. In terms of the beauty of things, I get that by walking in the mountains every Saturday and looking at birds, trees, waterfalls, flowers, clouds, the sea. In terms of religious experience, it is what many Quakers have found in the gathered Meeting for Worship. We can appreciate and take all of this seriously.

7.2 By their fruits ye shall know them: the ultimate issue

And in the end there is a great power that emerges from that silence, as is shown by Quaker history in relation to prisons, peace, poverty, slavery, and so on, that is the ultimate proof there is something powerful going on in the silence of the Meeting for Worship. Here we meet faith not just as an intellectual enterprise but rather as an overall way of living, with an intellectual basis but with an emotional and value-laden dimension; on the foundation of that understood faith and those values, working to positively change the world around us in the hope that this will make a real difference for the good.

The viewpoint presented by Richard Dawkins and his colleagues simply does not begin to approach that powerful relevance to real-world issues of justice, peace, and wellbeing. What movement have they led in relation to poverty or slavery or ill health or any other aspect of improvement of quality life, of the relief of suffering?

In the end, it is by their fruits that ye shall know them. The scientific atheism being offered us provides a very barren and bleak prospect for the future—it does not appear to result in good works on the ground or love in the community. This fact alone shows how lacking in depth that position is. In contrast, we have the dictum of George Fox, that he and many Quakers have made a reality in their lives:

Walk cheerfully over the world answering that of God in every one.[76]

Put in another way, the essence is

> *You are confused about what has gone wrong, and how to set it right? Then listen. This is what Yahweh asks of you, only this: to act justly, to love tenderly, and to walk humbly with your God.*[77]

This is the guide we need, rather than any amount of academic argument.

7.3 Paradox and deep reality

The result is paradoxical: actions and understanding that do not make sense in the initial context can be suitable reactions in a transformed context. This is the nature of a *kenotic* worldview:

> The idea that success is achieved by not worrying about success intersects the notion that we find our lives by losing them. The notion that we must empty ourselves to serve as channels for the Tao is echoed in the life of Jesus—he who renounced all worldly power, he who emptied himself and became obedient unto death, even the death on a cross, so that God's power could be shown.[78]

The fundamental importance of this revelation is in terms of its transformation of how things are understood.

> Scarcity is a reflection of our inner condition, a condition in which we believe meaning will come about by clinging. The more we cling the more meaning recedes: no matter how many scarce things we have, we will always want more … In contrast to this grasping at life is the emphasis on letting go so central to all great spiritual traditions … at the heart of letting go is faith and trust … What is the paradox [of scarcity]? Simply that 'he who seeks his life shall lose it, but he who loses his life for my sake shall find it'. True abundance comes not to those set on securing wealth. But to those who are willing to share a life of apparent scarcity. Those who seek well-being, who grasp for more than their fair share, will find life pinched and fearful. They will reap only the anxiety of

needing more, and the fear that someday it will all be taken away ... grasping brings less, and letting go brings more. [79]

This all fits well with the Quaker view of the world and of how to behave.

7.4 The merit of doubt

The counterpart of faith and hope is doubt, and one might be tempted to regard it as an undiluted bad: for it undermines faith, does it not? But on a broader view, doubt is inevitable as a part of the overall package, because (as emphasised above) metaphysical uncertainty will always remain with us if we are honest about it. Indeed, it is a bad sign if we have no doubt, for then we are in the hold of the kind of certainty that is the hallmark of unquestioning fundamentalism. So doubt is a sign of a *kenotic* faith: one that gives up the demand of certainty, and makes a leap of trust, despite the uncertainties that inevitably exist. It is a mark of a mature faith.

7.5 Reprise: *kenosis* and transcendence again

The nature of what I am trying to say is beautifully expressed in this *kenotic* prayer, written by Alan Gijsbers[80]:

They call on me to empty myself, Lord,
After all, you did.

You left the splendour of heaven
For humiliation on earth
And bowed to a servant role;
A felon's death.

But I am nothing. Already empty,
I have nothing to give, but my own broken self,
Conscious of pride, selfishness and sin,
I want to do good but I don't,

I don't want to do evil, but I do.
I cannot follow your path;
It is too steep, I have no power,
I lack the faith that I can
Walk your way.

I have tried so many times,
And failed just as often.
Can you still accept me, a failure?

Will you out of your power
Give me strength, give me hope,
Give me faith?
Will you so fill me with your transforming love
That I too may have the courage
To follow on that journey of surrender and sacrifice?

May my eyes be open to your power and healing,
May I lift my view from my self to you.

May I lift my eyes from myself in my weakness,
And see the needs of others around me
And help to heal them, out of my brokenness, restored by you.

May I see the rich possibilities that your love can create,
In me, and in all I meet,
In this needy world I'm in.

May I then see how I can create the space of love for the other,
Just as you created this world,
And you created space for me.
May I live, giving them the freedom to develop
To be what they choose to be,

Without manipulation or coercion,
But with grace.

May I become an agent of hope and peace
Of love and faith and joy
For your glory's sake.

As to transcendence, I conclude with some music that has for me a transcendent quality. I hope it does for you too.

[Play music: *Precious Lord* by the Pro Cantu Youth Choir].

Endnotes

1 Richard Dawkins, *The God delusion*, Bantam Books, 2006.
2 Daniel Dennett, *Breaking the spell: religion as a natural phenomenon*, Penguin, 2007.
3 Victor J Stenger, *God: the failed hypothesis. How science shows that God does not exist*, Prometheus Books, 2007.
4 Stephen Gould refers to 'Non-overlapping magisteria' (NOMA), see *Rocks of ages*, Ballantine Books, 1999.
5 See, for example, Nancey Murphy & George Ellis, *On the moral nature of the universe*, Fortress Press, 1995; Francis Collins, *The language of God: a scientist presents evidence for belief*, Free Press, 2006; John Polkinghorn, *Exploring reality: the intertwining of science and religion*, Yale University Press, 2007, and references therein.
6 Peter Atkins, 'The limitless power of science', in John Cornwall (ed.), *Nature's imagination*, Oxford University Press, 1995.
7 See, for example, Edward Harrison, *Cosmology*, Cambridge University Press, 2000; Joseph Silk, *On the shores of the unknown: a short history of the universe*, Cambridge University Press, 2005.
8 Neil A Campbell, *Biology*, Benjamin Cummings, 1991.
9 The 'Intelligent Design' movement is a political rather than scientific project.
10 Martin Rees, *Our cosmic habitat*, Princeton University Press, 2001.
11 For more detailed discussions, see GFR Ellis 'On the nature of emergent reality', in P Clayton and PCW Davies (eds), *The re-emergence of emergence*, Oxford University Press, 2006; 'Physics and the real world', *Foundations of Physics*, Apr 2006, 1–36, available at http://www.mth.uct.ac.za/~ellis/realworld.pdf .
12 Stafford Beer, *Brain of the firm*, Wiley, 1994.
13 Merlin Donald, *A mind so rare*, WW Norton, 2001.
14 October 2004 findings from The International Human Genome Sequencing Consortium.
15 Peter L Berger and Thomas Luckmann, *The social construction of reality*, Anchor Books, 1966.

16 Rational, adj. 1: (of behaviour, ideas, etc.)—based on reason rather than emotions: a *rational argument/ choice/ decision—rational analysis/ thought*. 2: (of a person) able to think clearly and make decisions based on reason rather than emotions. *Oxford advanced learner dictionary of current English*, Oxford University Press, 2000.
17 PW Atkins, 'The limitless power of science', in J Cornwell (ed.) *Nature's imagination: the frontiers of scientific vision*, Oxford University Press, 1995, pp. 122-32.
18 George FR Ellis, 'On rationality, emotion, faith, and hope: being human in the present age', in A Schutte (ed.), *Humanity in science and religion: the South African experience*, Cluster Publications, 2006.
19 RP Crease, 'The paradox of trust in science', *Physics World*, March 2004, p. 18.
20 David G Myers, *Intuition: its powers and perils*, Yale University Press, 2003.
21 Margaret Boden, *The creative mind: myths and mechanisms*, Abacus, 1994. Arnold H Modell, *Imagination and the meaningful brain*, MIT Press, 2003.
22 Robert Burns, 'To a mouse', http://www.robertburns.org/works/75.shtml.
23 Antonio Damasio, Descartes' error, Harper Collins, 2000; *The feeling of what happens*, Vintage, 2000.
24 This is memorably demonstrated in remarks made by Palmer Joss to Eleanor Arroway in the film *Contact*, directed by Robert Zemeckis, 1997.
25 George FR Ellis & Judith A Toronchuk, 'Neural development: affective and immune system influences', in Ralph D Ellis & Natika Newton (eds), *Consciousness and emotion*, John Benjamins, 2005, pp. 81–119.
26 Murphy & Ellis, *On the moral nature of the universe*.
27 Daniel Goleman, *Social Intelligence*, Arrow Books, 2007.
28 Jack Martin, Jeff Sugarman, & Janice Thompson, *Psychology and the question of agency*, State University of New York Press, 2003. Robert Kane, *A contemporary introduction to free will*, Oxford University Press, 2005.

29 John Barrow & Frank Tipler, *The anthropic cosmological principle*, Oxford University Press, 1988. Martin Rees, *Just six numbers*, Basic Books, 2000; *Our cosmic habitat*, Princeton University Press, 2001.
30 This is formalised in the idea of the 'landscape' of string theory—a vast array of possibilities for fundamental physics apparently allowed by the huge variety of possible vacua for that theory. See Leonard Susskind, *The cosmic landscape*, Little Brown, 2005.
31 Martin Rees, *Just six numbers*; Leonard Susskind, *The cosmic landscape*.
32 Alexander Vilenkin, *Many worlds in one: the search for other universes*, Hill & Wang, 2006.
33 George Ellis, 'Issues in the philosophy of cosmology', in J Butterfield and J Earman (eds), *Handbook in philosophy of physics*, Elsevier, 2006, pp. 1183-1285; available at http://arxiv.org/abs/astro-ph/0602280.
34 Martin Gardner, *Are universes thicker than blackberries?* Norton, 2003.
35 Murphy & Ellis, *On the moral nature of the universe*.
36 Viktor Frankel, *The doctor and the soul: from psychotherapy to logotherapy*, Souvenir Press, 2004.
37 Richard Weikart, *From Darwin to Hitler: evolutionary ethics, eugenics, and racism in Germany*, Palgrave Macmillan, 2004.
38 WD Hamilton, 'The evolution of altruistic behavior', *The American Naturalist*, 97, pp. 354-6, 1963. Robert Axelrod, *The evolution of cooperation*, Penguin, London, 1990. PJB Slater, 'Kinship and altruism', in PJB Slater and TR Halliday (eds), *Behaviour and evolution*, Cambridge University Press, Cambridge, 1994.
39 In 1940, the American Quaker AJ Muste stood up and ministered in a Quaker meeting, 'If I cannot love Hitler, I cannot love at all'.
40 J Neusner and B Chilton (eds), *Altruism in world religions*, Georgetown University Press, 2005.
41 Stephen G Post, *Unlimited love: altruism, compassion, and service*, Templeton Foundation Press, 2003.
42 Jacques Monod made this claim in his book, *Chance and necessity*, Vintage, 1972.
43 GFR Ellis, 'True complexity and its associated ontology', in JD Barrow

et al. (eds), *Science and ultimate reality: quantum theory, cosmology and complexity*, Cambridge University Press, 2004.
44 Ellis & Murphy, *On the moral nature of the universe*.
45 Dawkins, *The God delusion*.
46 Stenger, *God: the failed hypothesis*.
47 Raimond Gaita, *Good and evil: an absolute conception*, Routledge, 2004.
48 We no longer have public torture in London nor burnings in the streets of Oxford, for example; and slavery is no longer acceptable in most of the world (even though it takes place in some countries).
49 GR Ellis, 'The theology of the anthropic principle', in RJ Russell, N Murphy & CJ Isham (eds), *Quantum cosmology and the laws of nature*, Vatican Observatory, 1993, pp. 367-406. Murphy & Ellis. *On the moral nature of the universe*.
50 Raimond Gaita, *A common humanity: thinking about love and truth and justice*, Routledge, 2002.
51 William Temple, *Readings in St John's Gospel* (xxx), pp. xxix-xxxii.
52 Parker Palmer, *The promise of paradox: a celebration of contradictions in the Christian life*, Ave Maria Press, Notre Dame, 1980, pp. 51, 52.
53 Sir John Templeton, *Agape love*, Templeton Foundation Press. One does not necessarily need to be 'religious' in order to be *kenotic*, but I suspect it helps.
54 A Neave Brayshaw, *Christian faith and practice*, 1921, # 606.
55 See http://www.philadelphia reflections.com/reflections.php?content=topics_php/japan_ and_philadelphia.php, and David Hinshaw, *Rufus Jones, Master Quaker*, Arno Press, 1951, p. 279ff.
56 Murphy & Ellis, *On the moral nature of the universe*.
57 Jonathan Sacks, *The dignity of difference: how to avoid the clash of civilisations*, Continuum, 2002.
58 Parker Palmer, *The promise of paradox*, pp. 35-6.
59 Robert Bellah, *Beyond belief: essays on religion in a post-traditionalist world*, University of California Press, 1991.
60 Parker Palmer, *The promise of paradox*, pp. 39-40.
61 For a splendid meditation on immanence and transcendence, see Gerard

W Hughes, *God in all things*, Hodder & Stoughton, 2003.

62 Thich Nhat Hanh, *The miracle of mindfulness*, Beacon Press, 1999.

63 Antoine de St Exupery, *Flight to Arras*, Penguin Books, 1995.

64 Rufus Jones, 'Where the beyond breaks through', *The Friend*, Vol. 60, new series, 1920, p. 26.

65 John Lane, *Timeless beauty in the arts and everyday life*, Green Books, 2003.

66 See, for example, Philip Sherrard, *The sacred in life and art*, Golgonooza Press, 1990.

67 WE Paden, *Interpreting the sacred: ways of viewing religion*, Beacon Press, 1992.

68 See Gordon S Wakefield (ed.), *A dictionary of Christian spirituality*, SCM Press, 1989.

69 Geoffrey Hubbard, *Quaker by convincement*, Penguin Books, 1976. George H Gorman, *The amazing fact of Quaker worship*, Penguin, 1993.

70 John 1:1-14.

71 George FR Ellis: 'The theology of the anthropic principle', in RJ Russell, N Murphy & CJ Isham (eds), *Quantum cosmology and the laws of nature*, Vatican Observatory, 1993, pp. 367-406. Murphy & Ellis, *On the moral nature of the universe*.

72 J Neusner & B Chilton (eds), *Altruism in world religions*, Georgetown University Press, 2005.

73 I have been given the comment that people with different worldviews may not necessarily agree with this description of what constitutes moral reality and how people's actions are thereby determined. For instance, the Indigenous Australian worldview (The Dreaming) is a holistic integration of physical reality, spiritual reality and the Law (the foundation of human behaviour) and still provides for many people a very satisfying basis for moral behaviour which is quite different from what is presented here. That does not necessarily undermine my view of the deep nature of moral reality; different peoples may perceive it differently.

74 I have always liked the vision of afterlife given by CS Lewis in *The great divorce*.

75 Walter Wink, *The powers that be*, Doubleday, 1998. See http://www.westarinstitute.org/Periodicals/4R_Articles/Wink_bio/wink_bio.html.
76 From the journal of George Fox, p. 263—Fox's letter to ministers (1656) from prison in Launceston as written down by Ann Downer. See http://www.pym.org/pm/comments.php?id=67_0_49_0_C for the full text.
77 Micah 6:8, Jerusalem Bible translation.
78 Parker Palmer, *The promise of paradox*, p. 36.
79 Parker Palmer, *The promise of paradox*, pp. 99-101
80 Personal communication from Alan Gijsbers, of Melbourne, to author, July 2005.

THE JAMES BACKHOUSE LECTURES

The lectures were instituted by Australia Yearly Meeting of the Religious Society of Friends (Quakers) on its establishment in 1964.

They are named after James Backhouse who, with his companion, George Washington Walker, visited Australia from 1832 to 1838. They travelled widely, but spent most of their time in Tasmania. It was through their visit that Quaker Meetings were first established in Australia.

Coming to Australia under a concern for the conditions of convicts, the two men had access to people with authority in the young colonies, and with influence in Britain, both in Parliament and in the social reform movement. In meticulous reports and personal letters, they made practical suggestions and urged legislative action on penal reform, on the rum trade, and on land rights and the treatment of Aborigines.

James Backhouse was a general naturalist and a botanist. He made careful observations and published full accounts of what he saw, in addition to encouraging Friends in the colonies and following the deep concern that had brought him to Australia.

Australian Friends hope that this series of Lectures will bring fresh insights into the Truth, and speak to the needs and aspirations of Australian Quakerism. This particular lecture was delivered at Monash University, Melbourne, Victoria, on Monday, 7 January 2008, during the annual meeting of the Society.

Lyndsay Farrall
Presiding Clerk
Australia Yearly Meeting

THE **JAMES BACKHOUSE** LECTURES

1990   *Quakers in Politics: Pragmatism or Principle?*
Jo Vallentine & Peter Jones

1991   *Loving the Distances Between: Racism, Culture and Spirituality*,
David James & Jillian Wychel

1993   *Living the way: Quaker Spirituality and Community*, UJ O'Shea

1994   *As the Mirror Burns: Making a Film about Vietnam*, Di Bretherton

1995   *Emerging Currents in the Asia-Pacific*, DK Anderson & BB Bird

1996   *Our Children, Our Partners – a New Vision for Social Action in the 21st Century*, Elise Boulding

1997   *Learning of One Another: The Quaker Encounter with Other Cultures and Religions*, Richard G Meredith

1998   *Embraced by Other Selves: Enriching Personal Nature through Group Interaction*, Charles Stevenson

1999   *Myths and Stories, Truths and Lies*, Norman Talbot

2000   *To Learn a New Song: A Contribution to Real Reconciliation with the Earth and its Peoples*, Susannah Kay Brindle

2001   *Reconciling Opposites: Reflections on Peacemaking in South Africa*, Hendrik W van der Merwe

2002   *To Do Justly, and to Love Mercy: Learning from Quaker Service*, Mark Deasey

2003   *Respecting the Rights of Children and Young People: A New Perspective on Quaker Faith and Practice*, Helen Bayes

2004   *Growing Fruitful Friendship: A Garden Walk*, Ute Caspers

2005   *Peace is a Struggle*, David Johnson

2006   *One Heart and a Wrong Spirit: The Religious Society of Friends and Colonial Racism*, Polly O Walker

2007   *Support for Our True Selves: Nurturing the Space Where Leadings Flow*, Jenny Spinks

Backhouse Lectures, as well as other Australia Yearly Meeting publications, are available from Friends Book Sales, PO Box 181, Glen Osmond, South Australia 5064, Australia. Email <sales@quakers.org.au>.

www.ingramcontent.com/pod-product-compliance
Lightning Source LLC
Chambersburg PA
CBHW051715040426
42446CB00008B/892